SCIENCE CHALLENGE

ANTHONY D. FREDERICKS

Good Year Books
An Imprint of Addison-Wesley Educational Publishers, Inc.

Dedication

To Bobbie Dempsey, for her warm editorship and loyal friendship. May they always be constants!

Good Year Books are available for most basic curriculum subjects plus many enrichment areas. For more Good Year Books, contact your local bookseller or educational dealer. For a complete catalog with information about other Good Year Books, please write:

Good Year Books
1900 East Lake Avenue
Glenview, IL 60025

Cover illustration, part opener illustrations, and book design by Amy O'Brien Krupp.

All other illustrations by Dave Garbot.

Text copyright © 1998 Anthony D. Fredericks.

Illustrations © 1998 Good Year Books.

All Rights Reserved.

Printed in the United States of America.

ISBN 0-673-36373-2

1 2 3 4 5 6 7 8 9 - PK - 04 03 02 01 00 99 98 97

Only portions of this book intended for classroom use may be reproduced without permission in writing from the publisher.

CONTENTS

INTRODUCTION 1

DAILY PROBLEMS

    LIFE SCIENCE 7

    PHYSICAL SCIENCE 15

    EARTH SCIENCE 23

    SPACE SCIENCE 31

EXTENDED CHALLENGES ...39

ANSWER KEY 57

INTRODUCTION

All kids are fascinated with science. Each child has a natural curiosity about the world and an innate desire to learn more about things that have an impact on his or her daily life. ("Why is the sky blue?" "Where do babies come from?" "Why does my shadow follow me?") This natural curiosity can be stimulated in the elementary classroom through an active approach to science. In fact, students learn best when they take a participatory role in pursuing self-initiated questions and engaging in a "hands-on, minds-on" approach to science. *Science Challenge* is designed to help students use their scientific curiosity and knowledge in real-life explorations that expand and extend the science program in numerous ways.

NATIONAL SCIENCE EDUCATION STANDARDS

The National Science Education Standards have provided classroom teachers with a road map of what students need to know, understand, and be able to do at different grade levels in order to be scientifically literate. The Standards offer guidelines for the development and maintenance of viable and dynamic science programs. A "process approach to science" is emphasized through the Standards, and students are engaged in inquiry-based skills that emphasize critical and logical thinking. In short, students actively develop their understanding of science by combining scientific knowledge with reasoning and thinking skills.

Science Challenge is focused on the promotion and enhancement of

The Teaching Standards of National Science Education Standards place emphasis on:

1. understanding and responding to individual student's interests, strengths, experiences, and needs
2. selecting and adapting curriculum
3. focusing on student understanding and use of scientific knowledge, ideas, and inquiry processes
4. guiding students in active and extended scientific inquiry
5. providing opportunities for scientific discussion and debate among students
6. continuously assessing student understanding
7. sharing responsibility for learning with students
8. supporting a classroom community with cooperation, shared responsibility, and respect[1]

[1] "Changing Emphases: Science Teaching Standards."
National Science Education Standards. National Research Council. Washington, DC, 1996, p. 52.

the National Science Education Standards—specifically the Teaching Standards. As outlined by the National Research Council and the National Science Teacher's Association, those standards emphasize the concepts listed in the box on page 1. Each of these concepts is embedded within the activities, problems, and processes of this book.

Science Challenge helps you facilitate the promotion of those standards, while offering your students highly engaging activities that focus on science as a problem-solving experience. As a result, your students will begin to understand science as an important and dynamic part of their everyday lives, not just as a classroom subject.

THE PROBLEM-SOLVING APPROACH

This book provides problem-solving activities in which students can use higher level thinking skills together with basic scientific information. Too frequently, students are given piles of factual data but little opportunity to think through various situations, formulate opinions, justify their responses, or interact with their classmates. The activities in this book challenge students by stimulating them to move beyond rote memorization of facts into development of complex thoughts and personal discoveries. Here is a sample problem:

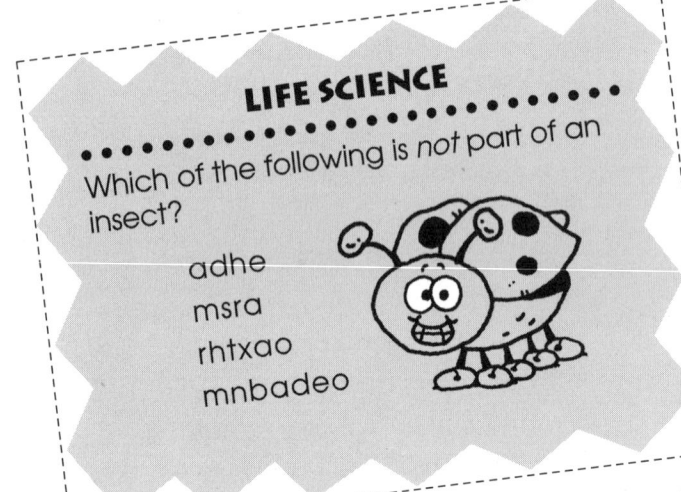

LIFE SCIENCE

Which of the following is *not* part of an insect?

adhe
msra
rhtxao
mnbadeo

Several skills are necessary to solve this problem:

1. The student must rearrange the letters in each word in the proper sequence (language arts, spelling). Note that, in puzzles throughout this book, two-word names are treated as one name with no space between.

2. The student must obtain information about insects to identify their body parts (encyclopedia, children's books).

3. The student must identify those body parts that belong to insects and those that do not (problem solving, critical thinking).

Science Challenge presents a collection of activities, in a stimulating and enjoyable format, that encourages students to actively process scientific information. Students will use science process skills such as *measuring*, *classifying*, *inferring*, *predicting*, *observing*, *experimenting*, and *communicating*. Intended for students in Grades 4 to 6, the book is also appropriate for gifted pupils in the lower grades.

CONTENT CATEGORIES

The activities and information in this book are organized in four different areas—Life Science, Physical Science, Earth Science, and Space Science.

1. LIFE SCIENCE

Life: It's all around us. From Rover barking in the backyard, to the plants in the living room, to the tiny speck of mold on the kitchen counter, we are surrounded by life. Understanding how plants and animals grow, develop, and interact with each other is an important part of science. In many ways, life is the area of the scientific world with which students are most familiar and is truly a field ripe for exploration. As students gain an awareness of the life forms around them, they also gain an appreciation for their own place in the gigantic ecosystem that we all participate in every day.

2. PHYSICAL SCIENCE

From the time we get up in the morning until we climb back into bed at night, our lives are influenced by a variety of scientific laws and principles. Although we may give little thought to the soap floating in the bathtub, the static electricity in the carpet, or the mechanical can opener on the kitchen counter, they are all governed by basic tenets of science. The need to understand the forces that regulate our lives underscores the importance of physical science.

3. EARTH SCIENCE

Four and a half billion: a number almost too large to comprehend. Yet, that's how many years the earth has been in existence. During that time, it has undergone some remarkable changes. Rocks have formed, primeval seas have ebbed and flowed across vast continents, and dramatic weather conditions have contributed to the geography and structure of our planet. The ground beneath our feet, its composition and design, and the forces that continue to shape it are magnificent and spectacular—and their study can be equally so within the science curriculum.

4. SPACE SCIENCE

It's amazing to realize that planet Earth is only a microcosm in the vastness of the universe. It's but one particle in a galaxy of stars, satellites, meteorites, and other celestial bodies. Humans have been constantly fascinated with what's "out there." Telescopes, observatories, and complex space probes have revealed some of the mysteries of the universe. They have also underscored the incredible amount of information we still need to learn. Given the media's emphasis on space exploration, it is one of the most intriguing areas for discovery within the elementary science program.

The data presented within these activities have been checked against many science texts normally used in the intermediate grades. These problems represent a cross section of the information commonly presented within each of the four sciences. Students thus have many opportunities to use their prior knowledge along with their problem-solving skills to work out appropriate responses.

DAILY PROBLEMS AND EXTENDED CHALLENGES

The activities in this book are organized into two groups—Daily Problems and Extended Challenges.

DAILY PROBLEMS

The book contains 192 Daily Problems—enough for every day of the school year. There are 48 problems in Life Science, 48 in Physical Science, 48 in Earth Science, and 48 in Space Science. Initially, you may wish to remove these pages from the book, duplicate them, paste them on oaktag, laminate them, and cut them into cards. Arrange the cards in a file box in sequential order or randomly. The problems can be used in one or more of the following ways:

1. When students arrive in the morning, or during a few minutes at the end of the day, ask them each to select a card at random and work on the listed problem. Personal charts can be initiated and individually maintained to record the problems each pupil solves.

2. Depending on the structure of your science text, you may wish to have students work in one area (Life, Physical, Earth, or Space) until most or all of the problems in that section have been solved. Students can then move on to another section.

3. Post one card on the bulletin board for all students to solve during their free time.

4. Have students work in pairs, exchanging ideas and working together toward a mutual solution. This technique is particularly appropriate for below-level students.

5. Assign one card per day per student as a homework assignment.

EXTENDED CHALLENGES

The Extended Challenges require long-term investigations by students. These pages are designed to challenge students in:

- assembling data
- interpreting facts
- organizing thoughts
- processing information

A word of caution is in order. Both the Daily Problems and the Extended Challenges are designed as reinforcing activities and are not intended for the initial learning of scientific information. Both kinds of activities are most appropriate as a follow-up to the data and concepts taught through your science curriculum. All these activities can be used to strengthen and promote important ideas enumerated in your science textbook. Thus they serve as a valuable adjunct to the entire science curriculum.

Students will need to use various types of reference materials, such as encyclopedias, children's books, technological resources, and the like to solve the problems successfully. You may wish to post an Extended Challenge on the bulletin board at the beginning of the week and provide multiple opportunities for students, either individually or in groups, to complete the page. These challenges can be used as homework assignments, too. However you plan to use these pages, it will be important for you to plan some time to discuss and share students' findings.

Although these activities are designed to enhance and extend your science program, you should also encourage students to create similar problems and challenges for their classmates. This type of activity promotes the concept of active participation and stimulates a participatory approach to the mastery of scientific concepts. Student involvement in designing other activities makes the study of science exciting and dynamic. In turn, students are motivated to learn more about the world in which they live.

These activities have been formulated for use in a variety of classrooms and a host of learning situations. You can use them as an adjunct to the science curriculum or as a reinforcement tool for vital concepts. Additionally, you are encouraged to use them in individual, small-group, large-group, or whole-class instructional situations. You can use the activities in whatever sequence you feel most appropriate. These problems and challenges offer you many opportunities to "energize" science for your entire class.

THE ANSWER KEY

Answers to both the Daily Problems and Extended Challenges are presented at the end of this book. For some of the Extended Challenges, potential solutions are offered because some answers depend on students' background and experience and on the challenging nature of the world. When students' answers differ in some respect from those provided in the answer section, plan time to discuss students' rationale and reasoning. You may need to consult current resources (newspapers, magazines, technological sources) to verify and/or confirm some responses. Helping students understand that science is not a static subject will be an important by-product of working with this book.

Daily Problems
LIFE SCIENCE

CARDS 1–48

From Science Challenge by Anthony D. Fredericks. Copyright © 1998 Good Year Books.

LIFE SCIENCE — 1

Which of the following is the largest of all human organs?

nksi   ctmhots   rhtae   veril

LIFE SCIENCE — 2

Tanisha ties a swing to a branch of an oak tree. The seat of the swing is 60 cm from the ground. If the tree grows 20 cm a year, how far off the ground will the swing be in 4 1/2 years?

LIFE SCIENCE — 3

In what part of the body would you find the retina, the cornea, and the iris?

LIFE SCIENCE — 4

Carrie, 14 years old, broke three bones in her right arm and two ribs when she was in a car accident. How many bones in her body remain unbroken?

LIFE SCIENCE — 5

Fill in the puzzle with the names of small groups of animals. (For example, fish often travel in a *school*.)

LIFE SCIENCE — 6

Which of the following is *not* a mollusk?

alnis   iqsdu   sroetbl   pocstou

LIFE SCIENCE — 7

Carnivorous plants are able to do something no other plants can do. What is it?

LIFE SCIENCE — 8

For a long time people referred to me as a **Brontosaurus**. But now I have a new name. What am I called now?

LIFE SCIENCE 9

Fill in the puzzle with the names of conifers.

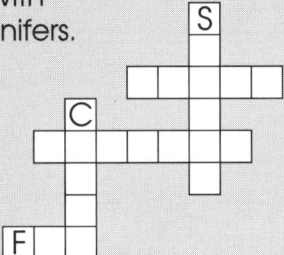

LIFE SCIENCE 10

I am a migrating bird. In fact, each year I migrate the longest distance of any animal. What am I?

LIFE SCIENCE 11

Which of the following is not a method of pollination?

rsdib   cisstne   nwdi   gthnsuil

LIFE SCIENCE 12

I am a group of living things. I cannot move or make my own food. I must get my food from living or dead plants and animals. What am I?

LIFE SCIENCE 13

For dinner, Tyrone ate a carbohydrate that is principally grown in both Idaho and Maine. What did he eat?

LIFE SCIENCE 14

Put a **T** in front of each true statement and an **F** in front of each false statement.

____ All mammals are carnivores.
____ A bear is an omnivore.
____ Herbivores only live on land.
____ Carnivores can eat both omnivores and herbivores.

LIFE SCIENCE 15

Which organism is incorrectly placed in the following food chain?

sunflower seeds → mouse → deer → red-tail hawk

LIFE SCIENCE 16

Match the following by drawing lines from the items on the left to the corresponding items on the right.

taproot	peanut
legume	carrot
monocot	bean
dicot	corn

LIFE SCIENCE — 17

Which of the following is *not* part of an insect?

adhe
msra
rhtxao
mnbadeo

LIFE SCIENCE — 18

Which of the following do *not* go through metamorphosis?

frog	butterfly
locust	marmot
moth	toad
limpet	shrimp

LIFE SCIENCE — 19

Place the following words in the correct order—from largest group to smallest group.

family, kingdom, genus, phylum, class, species, order

LIFE SCIENCE — 20

Fill in the puzzle with the names of parts of a flower.

LIFE SCIENCE — 21

Many people want to decrease their intake of $C_6H_{12}O_6$. What types of foods should they reduce?

LIFE SCIENCE — 22

I am the only North American marsupial. What am I?

LIFE SCIENCE — 23

A spider is holding its prey with its two front legs. How many legs is it not using?

LIFE SCIENCE — 24

Match the following by drawing lines from the trees on the left to the states they live in on the right.

sequoia	Colorado
cypress	California
aspen	Arizona
cottonwood	Florida

LIFE SCIENCE — 25

Which of these statements are *not* always true?

A. Vertebrates have a nervous system.
B. Vertebrates have backbones.
C. Vertebrates stand up straight.
D. Vertebrates are omnivorous.

LIFE SCIENCE — 26

In which of the following countries would you *not* find a poison-arrow frog?

ruep   npaaam
zraibl   naaadc

LIFE SCIENCE — 27

Put a **T** in front of each true statement and an **F** in front of each false statement about the Tyrannosaurus rex.

___ Lived during the Jurassic period
___ Lived in South America and South Africa
___ Stood approximately 18 meters tall
___ Was carnivorous

LIFE SCIENCE — 28

What do the following animals have in common?

gila monster
sidewinder
horned toad
rattlesnake

LIFE SCIENCE — 29

This plant lives in the tropics and is sometimes known as the "walking plant." What is it?

LIFE SCIENCE — 30

Fill in the puzzle with the names of the stages in the growth of a frog.

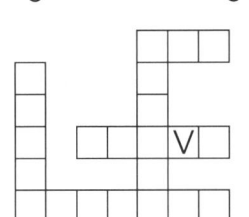

LIFE SCIENCE — 31

I am an animal that has very large incisors and cuspids. What kind of food do I eat?

LIFE SCIENCE — 32

Abdomen is to insect as _____ is to mammal.

_____ is to frog as ear is to dog.

Rattlesnake is to _____ as dog is to incisors.

Scales are to reptiles as feathers are to _____.

LIFE SCIENCE 33

What is wrong with the following sentences?
- A. Polar bears and penguins can be found across the continent of Antarctica.
- B. Camels and rattlesnakes can be found across the Sahara Desert.
- C. Lions and Asian elephants can be found across the country of Kenya.

LIFE SCIENCE 34

Match the following by drawing lines from the items on the left to their corresponding items on the right.

saguaro cactus   84 meters tall
sequoia tree   60 meters long
liana vines   15 meters tall
giant kelp   300 meters long

LIFE SCIENCE 35

Richard lives in the country where the only mammal that lays eggs lives. What country does he live in?

LIFE SCIENCE 36

The following animals all have something in common. What is it?

passenger pigeon
great Auk
dodo
plains grizzly bear

LIFE SCIENCE 37

Maria is using a scientific instrument to examine platelets. What type of instrument is she using?

LIFE SCIENCE 38

All of the following terms have something in common. What is it?

anvil
cochlea
stirrup
hammer

LIFE SCIENCE 39

Which of the following live in a tide pool?

mtilep   rskah   elgaa   ntau

LIFE SCIENCE 40

Which of the following are *always* true about eggs?
- A. They have shells
- B. They have a yolk.
- C. They are fertilized by males.
- D. They develop inside the female.

LIFE SCIENCE 41

Andrea wanted to interview a person who specialized in the study of insects. What type of scientist did she want to talk to?

LIFE SCIENCE 42

A bird stands up and looks around. Every direction it looks is north. What bird could it be and where is it?

LIFE SCIENCE 43

Fill in the puzzle with the names of arthropods.

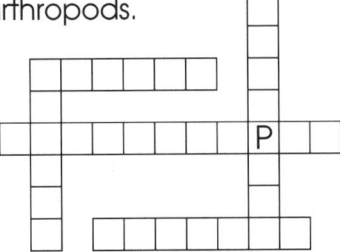

LIFE SCIENCE 44

_____ is to sheep as chick is to hen.

Quail is to _____ as whale is to pod.

Tigress is to tiger as _____ is to pig.

Cow is to barn as hare is to _____.

LIFE SCIENCE 45

Kai lives near an ecosystem that is often referred to as the "River of Grass." In what state does Kai live in?

LIFE SCIENCE 46

Which of the following is *not* an endangered environment?

ldwtenas   frrtseiaon

nxgoey   mtsraes

LIFE SCIENCE 47

Carson plants three bean seeds in three different pots. After two weeks the first grows to a height of 9 cm, the second to a height of 14 cm, and the third to a height of 20 cm. Which of the following could be true?

A. The soil in each pot was at a different temperature.
B. The seeds were planted at different depths.
C. The seeds were given different amounts of water.

LIFE SCIENCE 48

Which of the following prehistoric creatures was aquatic?

Plesiosaurus

Allosaurus

Iguanodon

Ichthyosaurus

Daily Problems
PHYSICAL SCIENCE

CARDS 49–96

PHYSICAL SCIENCE — 49

Which of these materials are conductors of electricity?

cpatlsi   cmcreai
rweat   prceop

PHYSICAL SCIENCE — 50

Martina wants to magnify a butterfly so that she can see the colorful scales on its wings. What type of lens should she use?

PHYSICAL SCIENCE — 51

Which of these will produce static electricity?

A. Rubbing an inflated balloon over a cotton shirt
B. Rubbing a glass rod with a wool cloth
C. Rubbing your hands together

PHYSICAL SCIENCE — 52

Water is a

A. mixture.
B. compound.
C. element.
D. polymer.

PHYSICAL SCIENCE — 53

Which of the following words is *not* an example of a simple machine?

finek   norbodok
esewas   doari

PHYSICAL SCIENCE — 54

Which of the following are chemical symbols for elements that burn?

He   S   Sn   C

PHYSICAL SCIENCE — 55

Paulette put four chlorine tablets into her swimming pool. Which two terms could be used to describe the water?

solvent   element
dilute   concentrated
solute   formula

PHYSICAL SCIENCE — 56

Fill in the following puzzle with words related to heat.

RADIATION

PHYSICAL SCIENCE 57

Match the following by drawing lines from the terms on the left to their corresponding terms on the right.

electric energy   airplane radio
potential energy   airplane
mechanical energy   airplane flying
kinetic energy   airplane fan

PHYSICAL SCIENCE 58

Put a **T** in front of each true statement and an **F** in front of each false statement.

____ Coal is a nonrenewable nuclear power source.
____ Gasoline is derived from natural gas.
____ Crude oil is a fossil fuel.
____ Uranium is a renewable fossil fuel.

PHYSICAL SCIENCE 59

In which of the following countries would you generate the least amount of solar energy during an entire year?

eocmxi   yknae
aiautsrlai   mvtaein

PHYSICAL SCIENCE 60

I make plant roots grow down. I keep things from falling off the Earth. I make balls bounce. What am I?

PHYSICAL SCIENCE 61

Boyd purchased 1600 grams of peanuts. How many pounds of peanuts did he have?

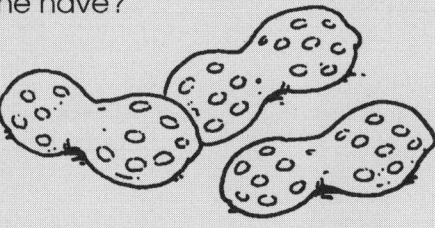

PHYSICAL SCIENCE 62

Jackson is wearing an outfit that absorbs all of the light that strikes it. What color is Jackson's outfit?

PHYSICAL SCIENCE 63

Which of these rates of speed is the fastest?

A. 28 miles per hour
B. 40 kilometers per hour
C. 19 knots per hour

PHYSICAL SCIENCE 64

Andi, Carole, and Michelle are skiing down a snow-covered mountain. Which of the following will eventually stop the forward motion of their skiis?

tcrfnioi   vyrgtia   iiaten

17

PHYSICAL SCIENCE — 65

Matilda poured herself a cup of hot tea and then let it cool off. What was happening to the particles of matter in the cup as the water cooled?

PHYSICAL SCIENCE — 66

Which of the following terms tells what gold is?

telneme   nopdumoc
mota   lelcomue

PHYSICAL SCIENCE — 67

What is the chemical formula for the chemical compound that covers more than two-thirds of the Earth's surface?

PHYSICAL SCIENCE — 68

I do not have shape or volume. You cannot see me, but I fill any container in which I am placed. I have weight, but you cannot hold me. What am I?

PHYSICAL SCIENCE — 69

Which of the following vegetables could be considered a wedge?

sbnae   trcrao   ncro   aspe

PHYSICAL SCIENCE — 70

Seymour wants to measure the force of gravity on a book. What instrument should he use?

PHYSICAL SCIENCE — 71

Fill in the following puzzle with words related to sound.

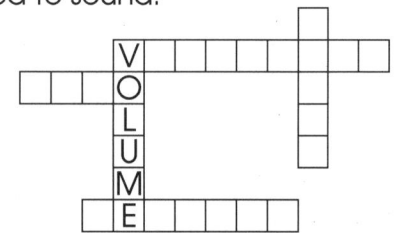

PHYSICAL SCIENCE — 72

Larry and Lori are sitting on a seesaw. Larry weighs 150 kg and Lori weighs 120 kg. If the seesaw is evenly balanced, who is sitting closer to the fulcrum?

PHYSICAL SCIENCE 73

How much farther does sound travel through water than through air in one second?

PHYSICAL SCIENCE 74

Tad is walking along a trail in the Grand Canyon. He shouts at a canyon wall that is 1661 meters away. How long will it take for the sound to come back to him as an echo?

PHYSICAL SCIENCE 75

Fiona is cooking steaks on her barbecue. What would be two examples of chemical changes that might be taking place on or in that barbecue grill?

PHYSICAL SCIENCE 76

Light energy is to lamp as _____ energy is to gasoline.

Sound energy is to horn as _____ energy is to radio.

Mechanical energy is to pistons as _____ energy is to toaster.

PHYSICAL SCIENCE 77

What is the name of a percussion instrument that produces a low-pitched sound?

PHYSICAL SCIENCE 78

Peter has a special instrument that vibrates 75,000 times per second. What animals can hear sound at that frequency?

PHYSICAL SCIENCE 79

Which of these are examples of levers?

A. A pair of scissors cutting a piece of paper
B. A batter hitting a baseball
C. A girl flying a kite
D. A wheelbarrow carrying a load of dirt

PHYSICAL SCIENCE 80

Match the following by drawing lines from the terms on the left to the corresponding terms on the right.

wheel and axle   snow shovel
wedge   stairs
inclined plane   doorknob
lever   knife

PHYSICAL SCIENCE 81

What is the mass of one liter of H_2O on the moon?

PHYSICAL SCIENCE 82

Which of the following words is another term for iron oxide?

tonegrin letse
goynex   sutr

PHYSICAL SCIENCE 83

Fill in the following puzzle with the names of conductors of electricity.

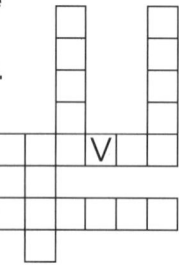

PHYSICAL SCIENCE 84

What is another name for an inclined plane wrapped around a post?

PHYSICAL SCIENCE 85

Which would cover the longer distance: light traveling for 5 seconds or sound traveling for 200 seconds?

PHYSICAL SCIENCE 86

Bernard visits the ophthamologist to have his eyes examined. She determines that Bernard is near-sighted. What shape will the lenses in his eyeglasses be?

PHYSICAL SCIENCE 87

I am used to measure current electricity. What am I?

PHYSICAL SCIENCE 88

Cynthia watches a thunderstorm on the horizon. She sees a bolt of lightning strike the ground. Four seconds later, she hears a clap of thunder. Approximately how far away is Cynthia from the storm?

PHYSICAL SCIENCE 89

Match the following by drawing lines from the items on the left to the corresponding items on the right.

potential energy   spring
kinetic energy   battery
                              wood
                              wood burning

PHYSICAL SCIENCE 90

Match the following by drawing lines from the items on the left to the corresponding items on the right.

positive charge   electron
neutral charge   proton
negative charge   neutron

PHYSICAL SCIENCE 91

Max places three jars on the table—a one-liter jar, a two-liter jar, and a four-liter jar. He fills each with boiling water. Which of the three jars produces the most heat?

PHYSICAL SCIENCE 92

Put a **T** in front of each true statement and an **F** in front of each false statement.

____ Work is equal to force x distance.
____ One joule equals one newton-meter.
____ Work is expressed in joules.
____ The unit for force is the newton.

PHYSICAL SCIENCE 93

Which of the following does *not* belong with the other three items?

    uranium   natural gas
    coal   crude oil

PHYSICAL SCIENCE 94

A perpetual motion machine is one that would never stop working once it was set in motion. Such a machine would not need an outside source of energy. It could produce its own energy forever. Such a machine cannot and does not exist because of one force of nature. What is that force?

PHYSICAL SCIENCE 95

Arrange the following elements from least number of protons to most number of protons.

    fsuulr   oinr
    bncora   muaclic

PHYSICAL SCIENCE 96

I am a material that does not conduct heat well. I am used in clothing and in houses. What am I?

Daily Problems
EARTH SCIENCE

CARDS 97–144

From *Science Challenge* by Anthony D. Fredericks. Copyright © 1998 Good Year Books.

EARTH SCIENCE 97

Sandra sets off on a raft at Seattle, Washington. If her raft drifts freely on the currents in the Pacific Ocean for 2800 kilometers, where will she end up?

EARTH SCIENCE 98

Merlissa lives in a city that has air pressure of about 15 pounds per square inch. Which of these cities does she not live in?

nsaodgei   rdneev
wenkryo   gciohac

EARTH SCIENCE 99

A dingo dog is watching water swirl down a bathtub drain in a clockwise motion. What country is the dog in?

EARTH SCIENCE 100

Fill in the puzzle with the names of the processes that act upon rocks to change them.

EARTH SCIENCE 101

I grew in a cave, but I am not alive. I began as water and minerals, but now I am solid. I grow upwards, but I never see the sky. What am I?

EARTH SCIENCE 102

Which hemisphere contains most of the world's oceans?

EARTH SCIENCE 103

Which of the following words name fossil fuels?

lio   odow   sag   loca

EARTH SCIENCE 104

An open bottle of soda pop, a large tub of water, and a cup of tea were all on the same table. On which container was there the greatest amount of air pressure?

24

EARTH SCIENCE — 105

Which of the following is a true statement?
- A. Magma heats to form sedimentary rocks.
- B. Magma cools to form metamorphic rocks.
- C. Magma heats to form igneous rocks.
- D. Magma cools to form igneous rocks.

EARTH SCIENCE — 106

Which of the Earth's continents are completely surrounded by water?

EARTH SCIENCE — 107

Some of earth's most spectacular landforms have been, and are being, created by an invertebrate animal. What is the animal and what land form does it create?

EARTH SCIENCE — 108

Which ocean has an area greater than that of all the continents combined?

EARTH SCIENCE — 109

Alexa wants to measure the magnitude of earthquakes. What instrument would she use?

EARTH SCIENCE — 110

Malcom is two-thirds the way up to the top of Mt. McKinley in Alaska. Chin is at the top of Mt. Washington in New Hampshire. Who is experiencing less air pressure?

EARTH SCIENCE — 111

I am a large mass of slowly moving frozen water. I have caused great changes on the surface of the earth. What am I?

EARTH SCIENCE — 112

Carmen is watching a series of cumulus clouds forming on the horizon. What type of weather will she probably experience in the next few hours?

EARTH SCIENCE 113

Robert notes that one part of tomorrow's weather report predicts wind speeds of 9 knots. What instrument will be used to measure that?

EARTH SCIENCE 114

Match the following by drawing lines from the items on the left to the corresponding items on the right.

latitude	Atlantic Ocean
longitude	east and west
International Date Line	Pacific Ocean
Greenwich Mean Time	north and south

EARTH SCIENCE 115

Francisco lives near the deepest ocean trench in the world. What country does he live in?

EARTH SCIENCE 116

Charles would like to explore a continent that is a little over half the size of North America and nearly half the size of Africa. There are no human inhabitants native to this continent, but plenty of animals. What continent does Charles want to visit?

EARTH SCIENCE 117

Which of the following terms tell about what breaks down rocks through erosion?

tarwe   slatnp
lisnama   dniw

EARTH SCIENCE 118

What is the approximate distance from the top of Mt. Everest to the bottom of the Marianas Trench?

EARTH SCIENCE 119

Henry would like to dig a hole 592 kilometers deep. If he could, what layer of the Earth would he reach?

EARTH SCIENCE 120

Salvadore's birthday in June is on the one day of the year that has the longest period of daylight. What hemisphere does he live in?

EARTH SCIENCE 121

The "Ring of Fire" includes four U.S. states. Which of those states is the largest?

EARTH SCIENCE 122

Which of the following minerals would be least immune to chemical weathering?

eosmlntie   ruqzta
bmeral   negarti

EARTH SCIENCE 123

Soft is to hard as _____ is to diamond

Dull is to graphite as _____ is to galena.

Coal is to nonmagnetic as _____ is to magnetic.

EARTH SCIENCE 124

Fill in the puzzle with terms that describe the features of the ocean floor.

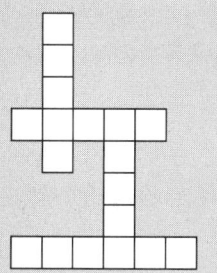

EARTH SCIENCE 125

Over a ten-year period, which of the following cities would receive the least amount of rainfall?

Manaus, Brazil
Antofagasta, Chile
Montevideo, Uruguay
Georgetown, Guyana

EARTH SCIENCE 126

Obsidian is formed when

A. lava reaches the surface, cools, and hardens slowly
B. lava reaches the surface, cools, and hardens quickly
C. lava stays underground and hardens slowly

EARTH SCIENCE 127

Which of the following is *not* a type of erosion?

ahet   crgalies   nwdi   twrae

EARTH SCIENCE 128

Tyree watches as scientists shoot an experimental rocket 60 kilometers into the air. What layer of the atmosphere will the rocket reach?

EARTH SCIENCE — 129

The Coriolis force makes winds

- A. spin
- B. die
- C. straighten
- D. curve

EARTH SCIENCE — 130

I am an air pollutant. Factories install scrubbers to reduce my release into the air when fossil fuels are burned. I stink. What am I?

EARTH SCIENCE — 131

The Mississippi delta is growing in size due to

- A. wave erosion
- B. wind erosion
- C. deposition
- D. glaciation

EARTH SCIENCE — 132

Fill in the puzzle with the names of metamorphic rocks.

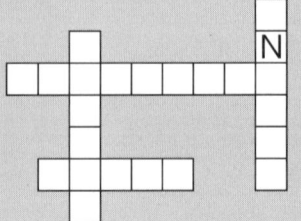

EARTH SCIENCE — 133

I am one of the world's most destructive storms. In Australia I am called a "willy-willy." In the Philippines I am known as a "baguio." What am I?

EARTH SCIENCE — 134

Which of the following islands was not created as the result of volcanic action?

nIgolidnsa
ahwiai
lidacne
arkaotka

EARTH SCIENCE — 135

Scientists recorded two separate earthquakes using the Richter scale. One earthquake measured 5.0 on the scale; the other released 30 times as much energy as the first. What did the second earthquake measure on the Richter scale.

EARTH SCIENCE — 136

Scientists tracked a tornado that was crossing Kansas. They gave it a classification of F-2. How much faster is the wind speed in a category F-2 tornado compared with a category 2 hurricane?

EARTH SCIENCE 137

Burning gasoline is to carbon monoxide as burning coal is to _____.

Noise pollution is to jet airplanes as _____ pollution is to car exhaust.

Thermal pollution is to heated water as water pollution is to _____.

EARTH SCIENCE 138

The Rocky Mountains are an example of _____ mountains.

lofd   obklc   dmoed

EARTH SCIENCE 139

The highest mountain in the world is how much taller than the highest mountain in Hawaii?

EARTH SCIENCE 140

Match the following by drawing lines from the items on the left to the corresponding items on the right.

polar climate   Barcelona, Spain
temperate climate   Portland, Oregon
tropical climate   Bombay, India
                      Great Bear Lake, Canada
                      Panama City, Panama

EARTH SCIENCE 141

For every square foot of the earth's surface, the atmosphere presses down with a weight of approximately _____.

EARTH SCIENCE 142

The South American (tectonic) plate includes all the following countries except one. Which one is it?

rpue   vbaiiol
nmaaap   lcihe

EARTH SCIENCE 143

Which of the following conditions is most likely to cause fog to form?

A. humid air, cool land
B. cool air, warm land
C. dry air, wet land
D. humid air, warm land

EARTH SCIENCE 144

What is wrong with the following experiment? Ramon obtains an empty glass jar. Using a ruler and a felt-tip marker, he marks the side of the jar in 2-centimeter increments. He places the jar outdoors. A week later he notes that there is 4 1/2 centimeters of water inside. He concludes that 4 1/2 centimeters of rain fell that week.

Daily Problems
SPACE SCIENCE

CARDS 145–192

SPACE SCIENCE 145

Julio weighs 240,000 grams on Earth. How much would he weigh on the moon?

SPACE SCIENCE 146

I hold the celestial record for the fastest revolution around this solar system's largest star. Who am I?

SPACE SCIENCE 147

I can travel where there are no molecules, but never through a brick. I can travel through space, but never around a corner. What am I?

SPACE SCIENCE 148

Fill in the following puzzle with the names of Jupiter's moons discovered by Galileo.

SPACE SCIENCE 149

Which of the following is *not* a unit of distance?

elgrtaiyh   tdmgaueni
casper   mkroleiet

SPACE SCIENCE 150

Match the following by drawing lines from the items on the left to the corresponding items on the right.

galaxy   Orion
constellation   Polaris
moon   Milky Way
star   Oberon

SPACE SCIENCE 151

Which of the following stars has the hottest surface temperature?

red star
blue-white star
yellow star

SPACE SCIENCE 152

Put a T in front of each true statement and an F in front of each false statement.

____ Gravity holds stars, gases, and dust together in a galaxy.

____ The brightness of a star is determined by its gases.

____ A star's magnitude depends on how close to the sun the star is.

SPACE SCIENCE 153

Match the following by drawing lines from the items on the left to the corresponding items on the right.

the Earth's core	2900 km thick
the Earth's crust	3550 km thick
the Earth's mantle	64 km thick

SPACE SCIENCE 154

How many stars form the bowl of the Big Dipper?

SPACE SCIENCE 155

Gary wanted to gather some information about the planet with the most moons. Which planet was he interested in?

SPACE SCIENCE 156

Which would have the greatest weight?

A. 44 grams of sugar on Earth
B. 2 kg of bananas on the moon
C. 1/2 kg of cereal on Earth

SPACE SCIENCE 157

The average distance from the sun of the planet Venus is 108,000,000 km. For Mars, the average distance is 228,000. How long would it take a beam of light to travel from Venus to Mars?

SPACE SCIENCE 158

Which of the following statements is not true?

A. There have been 12 manned explorations to the moon.
B. Erosion is responsible for the valleys on the moon.
C. The back side of the moon is covered with craters.
D. The face of the moon has large, flat plains.

SPACE SCIENCE 159

On which of the following planets would you weigh less than you do on Earth?

pentuen   tanrus
pejtiur   sevun

SPACE SCIENCE 160

Halley's Comet was seen twice in the 20th Century. Which sighting was closest to the bicentennial of the United States?

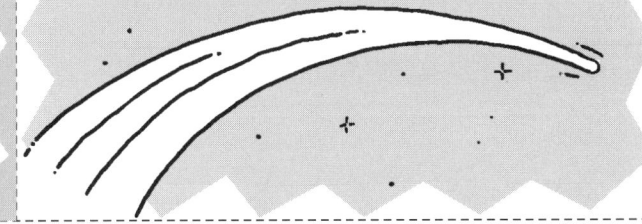

SPACE SCIENCE — 161

Which of the ringed planets is the smallest?

SPACE SCIENCE — 162

Fill in the puzzle with the names of the four gaseous planets

SPACE SCIENCE — 163

How long would it take a beam of light to travel from the Earth to the nearest star (not including the sun)?

SPACE SCIENCE — 164

Which of the following statements is *not* true about the planet Pluto?

A. It has two moons, Charon and Indrus.
B. It has a constant temperature of –230° C.
C. It has the largest orbit of any planet.
D. In 1983 it was the eighth planet from the sun.

SPACE SCIENCE — 165

Match the following by drawing lines from the items on the left to their corresponding items on the right.

the moon   light travels
Earth   sound travels
    light does not travel
    sound does not travel

SPACE SCIENCE — 166

Which of the following burns up in the Earth's atmosphere?

tocem
arts
ristadeo
roteem

SPACE SCIENCE — 167

Sirius is to star as Canis Major is to _____.

Milky Way is to _____ as Mars is to planet.

Cygnus is to summer sky as Orion is to _____.

SPACE SCIENCE — 168

Possible evidence of the existence of life forms has been discovered on which of the following:

A. Mars and Jupiter
B. Earth and Mercury
C. Mars and Earth
D. Earth and Neptune

SPACE SCIENCE — 169

Which of the following names something that includes all of the others?

lagyax   resdoita
sivrenue   ralos tssyme

SCIENCE — 170

If you lived on Jupiter, approximately how long would your summer vacation be, measured in Earth days?

SPACE SCIENCE — 171

How long would it take sound to travel 4 kilometers on the moon.

SPACE SCIENCE — 172

Fill in the puzzle with words related to the sun.

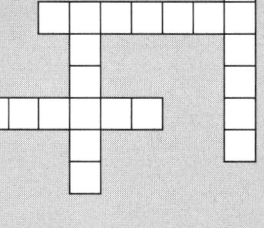

SPACE SCIENCE — 173

Three weeks after a new moon, how much of the moon will be lighted?

SPACE SCIENCE — 174

What takes the most time?

A. One rotation of the Earth
B. One revolution of the Earth
C. One rotation of the moon
D. One revolution of the moon

SPACE SCIENCE — 175

What is the approximate difference in temperature between the hottest planet and the coldest planet, measured in degrees Celsius?

SPACE SCIENCE — 176

Tamara is sailing her boat in the middle of the Pacific Ocean. She has lost all her maps and charts. What celestial body will help her find her way back home?

35

SPACE SCIENCE 177

The moons of Uranus were named differently than the moons of all the other planets. Where did the names of Uranus's moons come from?

SPACE SCIENCE 178

Colin lives in Christchurch, New Zealand. During what day of the year are the sun's rays most direct on that city?

SPACE SCIENCE 179

Uranus and Venus do something no other planets do. What is it?

SPACE SCIENCE 180

It is 36° C in Buenos Aires, Argentina. Which way is the Earth titled on its axis?

SPACE SCIENCE 181

A comet is moving away from the sun. Where is its tail?

A. Behind the comet
B. In front of the comet
C. It has no tail
D. Around the comet

SPACE SCIENCE 182

Put a **T** in front of each true statement and an **F** in front of each false statement.

____ An astronomical unit is equal to a light year.

____ A light year is more than an astronomical unit.

____ An astronomical unit is more than a light year.

SPACE SCIENCE 183

All of the features on the planet Venus are named for real or mythological women, except one. What is the name of that feature?

SPACE SCIENCE 184

Match the following by drawing lines from the items on the left to their corresponding items on the right.

Planet	Diameter
Neptune	6,790 km
Uranus	120,000 km
Saturn	48,600 km
Mars	51,138 km

SPACE SCIENCE 185

Sarah weighs 46 kg, Matthew weighs 53 kg, and Rica weighs 42 kg. What would be their combined weight on the planet Mercury?

SPACE SCIENCE 186

Arrange the following star colors from hottest to coolest.

yellow, red, blue-white, green, orange, blue

SPACE SCIENCE 187

Fill in the puzzle with the names of some of Saturn's moons.

SPACE SCIENCE 188

What is significant about the following period of time: 23 hours, 56 minutes, and 4.09 seconds.

SPACE SCIENCE 189

Where would you find the following geographic locations?

Tharsis Mountains, Olympus Mons, and Valles Marineris

SPACE SCIENCE 190

Which of the following is true?

A. The sun is the closest star in our galaxy.
B. The sun is the brightest star in our galaxy.
C. The sun is the only star in our galaxy.
D. The sun is the largest star in our galaxy.

SPACE SCIENCE 191

What is the only star that appears to stand still throughout the year?

SPACE SCIENCE 192

How many of the following planets have moons?

rnutrsa   tpoul   tarhe   nrsuua

Activity Sheets
EXTENDED CHALLENGES

1-16

Name: _____ Date: _____

EXTENDED CHALLENGE 1

Identify an example of each class of animals, and place each one in its proper habitat. One sample has been done for you.

CLASS	EXAMPLE	HABITAT
insect	iguana	savanna
	wolf	
fish	harp seal	river
	flamingo	
reptile	mallard	wetlands
	perch	
amphibian	frog	pond
	dragonfly	
bird	roadrunner	tundra
	moose	
mammal	dolphin	desert
	mosquito	
	alligator	rainforest
	buffalo	
	albacore	Everglades
	Florida panther	
	harpy eagle	ocean
	zebra	
	gazelle	prairie
	polar bear	
	bison	
	jackrabbit	
	sidewinder	
	salamander	

(insect — dragonfly — pond)

Name: _____   Date: _____

EXTENDED CHALLENGE 2

Place a check mark after each dinosaur to show which time period it lived in.

	TRIASSIC	JURASSIC	CRETACEOUS
Allosaurus	___	___	___
Ankylosaurus	___	___	___
Apatosaurus	___	___	___
Brachiosaurus	___	___	___
Coelophysis	___	___	___
Compsognathus	___	___	___
Iguanodon	___	___	___
Oviraptor	___	___	___
Psittacosaurus	___	___	___
Rutiodon	___	___	___
Stegosaurus	___	___	___
Triceratops	___	___	___
Tyrannosaurus	___	___	___

From Science Challenge by Anthony D. Fredericks. Copyright © 1998 Good Year Books.

41

Name: _____ Date: _____

EXTENDED CHALLENGE 3

Put a letter from the column on the left in front of item on the right to indicate which system each body part belongs to.

A. digestive
B. circulatory
C. respiratory
D. excretory
E. nervous
F. reproductive
G. muscular
H. skeletal
I. sensory

____ brain
____ tendon
____ alveoli
____ intestine
____ aorta
____ sperm
____ ureter
____ sweat gland
____ vein
____ tibia
____ nerves
____ ligaments
____ esophagus
____ trachea
____ ovary
____ kidney
____ triceps
____ salivary gland
____ diaphragm
____ eardrums
____ cartilage
____ pupils
____ valve
____ olfactory
____ axon

42

Name: _____   Date: _____

EXTENDED CHALLENGE 4

Complete the puzzle below by filling in the spaces with the names of endangered animal species.

Name: _____ Date: _____

EXTENDED CHALLENGE 5

Write YES or NO on each space following an activity to indicate what force(s) would be involved. The first one has been done for you.

	FRICTION	MAGNETISM	ELECTRICITY	GRAVITY
1. Ride a bicycle	yes	no	no	yes
2. Ski down a hill	___	___	___	___
3. Shine a flashlight	___	___	___	___
4. Operate a model train	___	___	___	___
5. Turn on a TV set	___	___	___	___
6. Push a car out of the snow	___	___	___	___
7. Walk on the moon	___	___	___	___
8. Make a telephone call	___	___	___	___
9. Brush your teeth	___	___	___	___
10. Use a computer	___	___	___	___

Name: _____ Date: _____

EXTENDED CHALLENGE 6

List all of the simple machines you can locate on the bicycle and on the can opener. One item has been done for you.

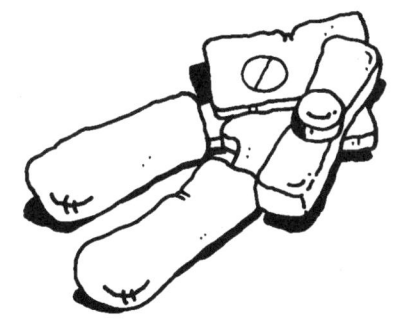

BICYCLE PART	SIMPLE MACHINE	CAN OPENER PART	SIMPLE MACHINE
wheels	wheel & axle		

Name: _____ Date: _____

EXTENDED CHALLENGE 7

Put a letter from the column on the left in front of each example of energy in the column on the right.

A. mechanical
B. thermal
C. chemical
D. electrical
E. nuclear

_____ gasoline
_____ battery
_____ light bulb
_____ atomic bomb
_____ bicycle
_____ oven
_____ computer
_____ generator
_____ pulley
_____ sun
_____ water wheel
_____ heating pad
_____ coal
_____ boiling water

Name: _____   Date: _____

EXTENDED CHALLENGE 8

Label each of the illustrations below with one of the following terms—**convection, radiation, conduction**—according to how heat is moving in each example.

1. _____

2. _____

3. _____

4. _____

5. _____

6. _____

47

Name: _____ Date: _____

EXTENDED CHALLENGE 9

In the puzzle below, locate and circle terms associated with different types of violent weather (including the names of storms). The terms will go down, across, or diagonally. One type of violent storm is missing from the puzzle. Write its name on the line under the puzzle. Two-part terms are spelled as one, with no space between parts.

```
W I N D C U R R E N T S S D Y
H R T U E L E C T R I C I T Y
U M W O X Q L P D L A S E B T
R W A V E S P I N I M Y E W H
R C B Y A O I W R G E T M N U
I Z C O P W S B M H A O E Y N
C T H U N D E R S T O R M S D
A R T Y U L V A E N P N V C E
N D F O E P E I M I B A B H R
E B L N R N P N R D D W Q H
P C N W O H J K L G E O W S E
P U E L D E P R E S S I O N A
F W C E X P L O S I O N W S D
P Y A S W H A I L S T O N E S
C M S T R O N G W I N D S S S
```

The name of the storm missing from the puzzle is: _____

48

Name: _____ Date: _____

EXTENDED CHALLENGE 10

For each of the mountain ranges or systems listed in the left column, indicate what type of mountain it is by placing a check mark on the space under the correct name. The first one has been done for you.

	FOLDED	FAULT-BLOCK	DOMED	VOLCANIC
1. Sierra Nevadas	___	X	___	___
2. Black Hills	___	___	___	___
3. Rocky Mountains	___	___	___	___
4. Appalachians	___	___	___	___
5. Cascades	___	___	___	___
6. Wasatch Range	___	___	___	___
7. Tetons	___	___	___	___
8. Ozarks	___	___	___	___

Name: _____ Date: _____

EXTENDED CHALLENGE 11

Which body of water is described by each set of clues?

1. This ocean surrounds the North Pole. It is frozen much of the year, and it contains many icebergs.

   _____

2. This is the largest body of water on Earth. It lies between Asia and North and South America.

   _____

3. This sea is north of South America. The Gulf Stream originates in its very warm waters.

   _____

4. This ocean is bordered by Saudi Arabia and Africa and was the location of many biblical stories.

   _____

5. Mostly located in the Southern Hemisphere, this body of water lies between Africa and Asia. It is the third largest ocean on Earth.

   _____

6. The United States and Mexico border this rich fishing area. Hurricanes occur frequently every year on this body of water.

   _____

7. Bordered by Europe and Africa, this is the largest enclosed sea on Earth. The first trade routes were established on this body of water many centuries ago.

   _____

8. This body of water is bordered by four provinces. It is the site of many early English explorations.

   _____

Name: _____ Date: _____

EXTENDED CHALLENGE 12

Fill in the spaces in the chart below with the correct names, terms, or descriptions. The first one has been done for you

KIND OF ROCK	MAJOR CLASSIFICATION	CHARACTERISTICS	HOW IT WAS FORMED
granite	igneous	large crystals	slowly cooling lava
shale		fine-grained	sediments of mud/clay/silt
		very hard; originally sandstone	great heat; great pressure
gneiss		salt & pepper appearance	
sandstone	sedimentary		
	igneous	gray, floats on water	rapidly cooling lava
diamond	metamorphic		
	metamorphic	white	
limestone	sedimentary		sediments of sand built up over time
		black, glassy	rapidly cooling lava

51

Name: _____ Date: _____

EXTENDED CHALLENGE 13

Some of the following statements are true about the planets, others are false. Identify those that are false, and rewrite them to make them correct.

_____ 1. Mercury, Venus, Mars, and Earth are called the inner planets.

_____ 2. Jupiter is 142,800 kilometers (88,700 miles) in diameter.

_____ 3. Triton is one of the moons of Neptune.

_____ 4. The symbol for Earth is ____.

_____ 5. Venus is the only planet named for a woman.

_____ 6. Saturn is not the only planet with rings.

_____ 7. Mars was named for the Roman god of war.

_____ 8. Mercury spins more slowly on its axis than Earth.

_____ 9. If you weigh 54 kg (100 lbs) on Earth, you would weigh 41 kg (97 lbs) on Uranus.

_____ 10. The temperature on Mercury ranges from 423°C (801°F) to –71°C (–279°F).

_____ 11. The original name for Uranus was "Herschel."

_____ 12. Saturn is referred to as the "Queen of the Planets."

_____ 13. Mars has two moons—Phobos and Deimos.

_____ 14. One of the planets has a symbol exactly like the biological symbol for "female."

_____ 15. It takes 247 Earth years for Pluto to travel around the sun.

Name: _____ Date: _____

EXTENDED CHALLENGE 14

Draw a line from each of the planets in the middle column to its average distance from the sun in the left column, and then to its corresponding length of day in the right column. One sample has been done for you.

AVERAGE DISTANCE FROM SUN (IN MILLIONS OF KM)	PLANET	LENGTH OF DAY (IN EARTH TIME)
4497	Mercury	116.7 days
1425	Venus	24.6 hours
228	Earth	10.4 hours
108	Mars	18.5 hours
5900	Jupiter	6.4 days
2867	Saturn	176 days
778	Uranus	24 hours
150	Neptune	9.9 hours
58	Pluto	16 hours

Name: _____ Date: _____

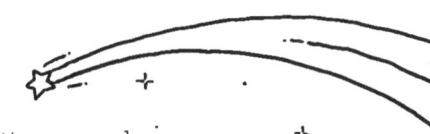

EXTENDED CHALLENGE 15

Write the missing information in the table of constellations and stars. The first one has been done for you.

CONSTELLATION	POPULAR NAME	BRIGHTEST STAR	BEST SEEN DURING
Taurus	Bull	Aldebaran	winter
	Chariot Driver	Capella	fall
Cepheus		none	all seasons
		Betelgeuse	winter
		Polaris	all seasons
Canis Major			
Cassiopeia			
Scorpius		Antares	
Leo	Lion		

Name: _____ Date: _____

EXTENDED CHALLENGE 16

Complete the puzzle below by filling in the spaces with the names of various celestial bodies. Treat multiple-word terms as one word, with no space between.

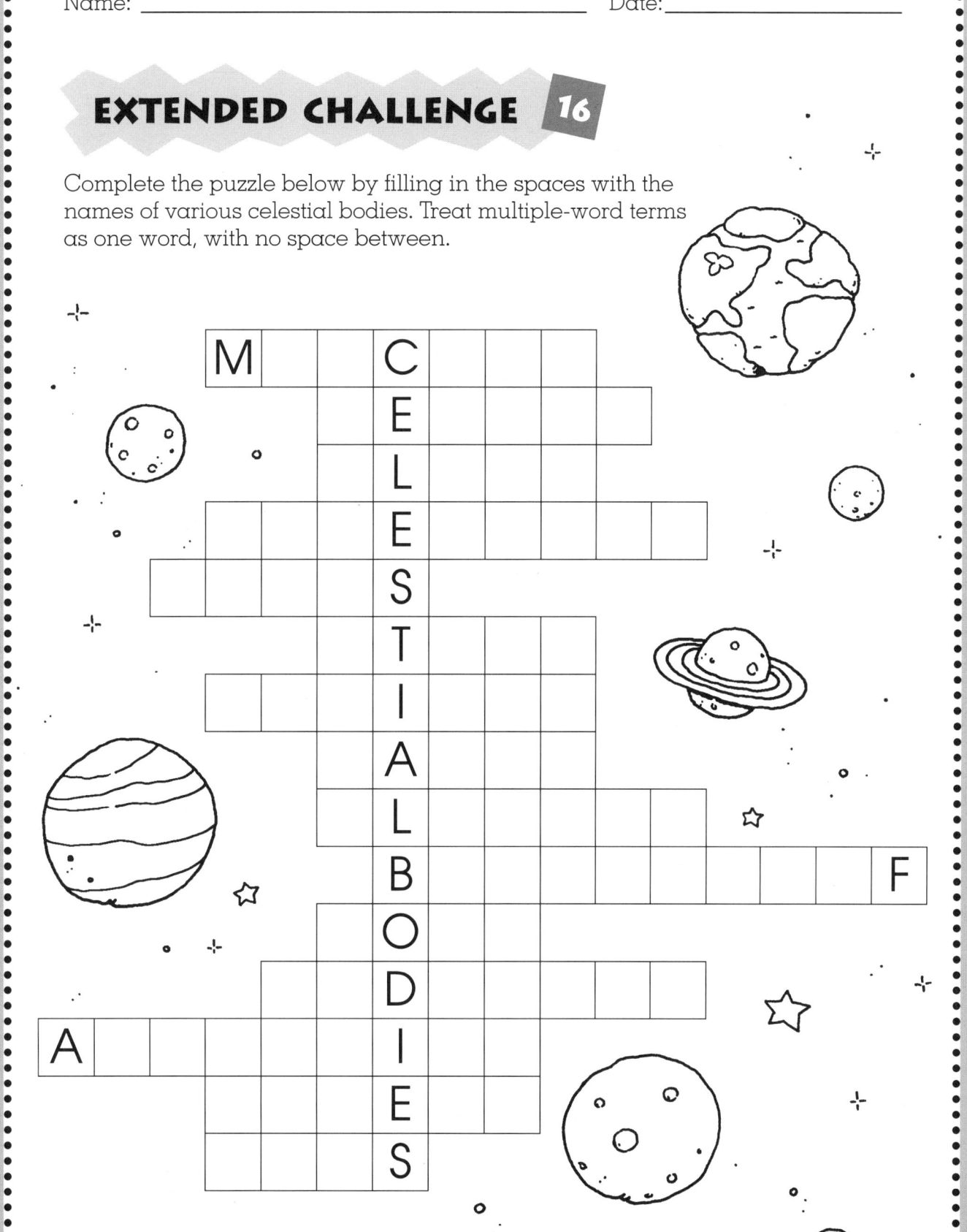

```
    M       C
            E
            L
            E
            S
            T
            I
            A
            L                        F
            B
            O
            D
            I
  A         E
            S
```

ANSWER KEY

ANSWER KEY

LIFE SCIENCE

1. (skin), stomach, heart, liver
2. 60 cm
3. the eye
4. 201
5.

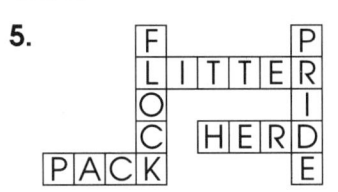

```
      F       P
  L I T T E R I
  O       H E R D
  C           I
  P A C K     E
```

6. snail, squid, (lobster), octopus
7. capture and digest animals
8. Apatosaurus
9.

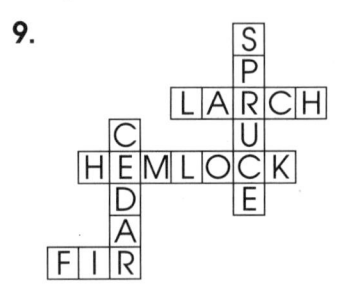

```
        S
        P
      L A R C H
    C   U
  H E M L O C K
    D   E
    A
    F I R
```

10. the arctic tern
11. birds, insects, wind, (sunlight)
12. fungi
13. potatoes
14. F, T, F, T
15. deer
16. taproot — peanut
    legume — carrot
    monocot — bean
    dicot — corn
17. head, (arms), thorax, abdomen
18. marmot, limpet, shrimp
19. kingdom, phylum, class, order, family, genus, species

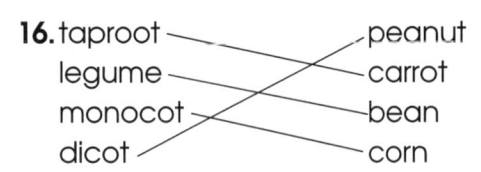

20.

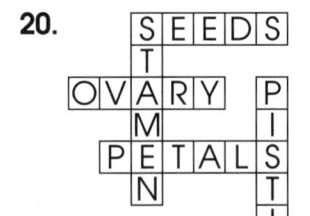

```
        S E E D S
        T
  O V A R Y   P
        M     I
      P E T A L S
        N     T
              I
              L
```

21. foods with sugar
22. the oppossum
23. 6
24. sequoia — Colorado
    cypress — California
    aspen — Arizona
    cottonwood — Florida
25. C & D
26. Peru, Panama, Brazil, (Canada)
27. F, F, F, T
28. all are desert reptiles
29. the mangrove tree
30.

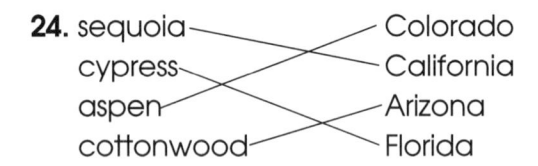

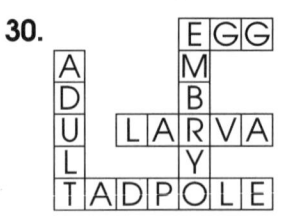

```
        E G G
        M
    A   B
    D   R
    U L A R V A
    L   Y
    T A D P O L E
```

31. meat
32. stomach, tympanum, whales, birds
33. A. Polar bears live in the arctic regions; penguins in Antartica.
    B. Camels live in the Sahara desert; rattlesnakes in deserts in the southwestern United States.
    C. Asian elephants live in India; lions live in African countries such as Kenya.
34. saguaro cactus — 84 meters tall
    sequoia tree — 60 meters long
    liana vines — 15 meters tall
    giant kelp — 300 meters long
35. Australia

ANSWER KEY

36. They are all extinct.
37. microscope
38. They are all parts of the ear.
39. (limpet), shark, (algae), tuna
40. None of the statements is *always* true.
41. entomologist
42. penguin, South Pole
43.

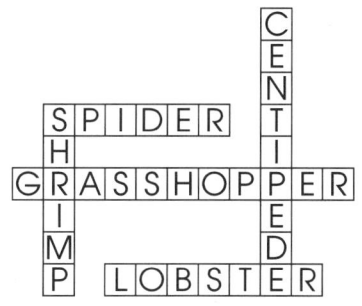

```
        C
        E
        N
   SPIDER
   H    I
   GRASSHOPPER
   R    E
   I    D
   MP LOBSTER
```

44. lamb, covey, sow, hutch
45. Florida
46. wetlands, rainforest, (oxygen), streams
47. All of them could be true.
48. Plesiosaurus, Ichthyosaurus

PHYSICAL SCIENCE

49. plastic, ceramic, (water), (copper)
50. convex lens
51. all three
52. 1 kilogram
53. knife, doorknob, seesaw, (radio)
54. (He = Helium), (S = Sulphur), Sn = Tin, (C = Carbon)
55. solvent, concentrated

56.

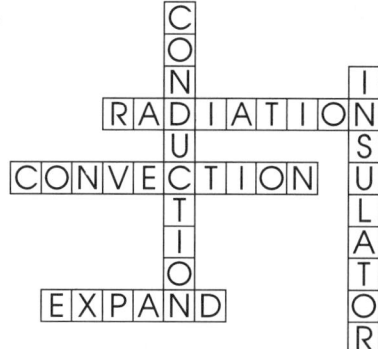

```
     C
     O
     N
     D
     U
  RADIATION   I
     T        N
  CONVECTION  S
     I        U
     O        L
              A
   EXPAND     T
              O
              R
```

57. electric energy —— airplane radio
    potential energy —— airplane
    mechanical energy —— airplane flying
    kinetic energy —— airplane fan
58. F, F, T, F
59. Mexico, Kenya, Australia, (Vietnam)
60. gravity
61. 3 1/2 pounds
62. black
63. A
64. (friction), gravity, inertia
65. They were slowing down.
66. (element), compound, atom, molecule
67. H$_2$O
68. oxygen/air/gas
69. beans, (carrot), corn, peas
70. scale
71.

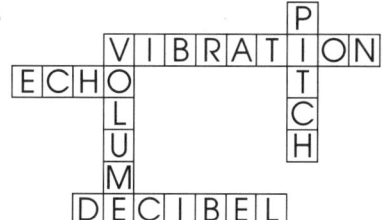

```
         P
    VIBRATION
  ECHO    I
    L     T
    U     C
    M     H
   DECIBEL
```

72. Larry
73. 1131 meters (3710 feet)
74. 5 seconds
75. burning charcoal, burning paint, rusting

ANSWER KEY

76. chemical, electrical, heat

77. bass drum

78. bat, porpoise

79. A, B, D

80. wheel and axle — snow shovel
wedge — stairs
inclined plane — doorknob
lever — knife

81. 1 kilogram

82. nitrogen, steel, oxygen, (rust)

83. crossword: STEEL, WATER, SILVER, IRON, COPPER

84. screw

85. light traveling for 5 seconds

86. concave

87. galvanometer

88. 1340 meters (4400 feet)

89. potential energy — spring, wood
kinetic energy — battery, wood burning

90. positive charge — proton
neutral charge — neutron
negative charge — electron

91. the four-liter jar

92. T, T, T, T

93. uranium

94. friction

95. carbon (6), calcium (20), sulfur (16), iron (26)

96. insulator

EARTH SCIENCE

97. Mexico

98. San Diego, (Denver), New York, Chicago

99. Australia

100. crossword: HEATING, PRESSURE, WEATHERING, COOLING

101. a stalagmite

102. Southern Hemisphere

103. (oil), wood, (gas), (coal)

104. The air pressure is the same on all three containers.

105. D

106. Australia, Antarctica

107. coral; coral reefs

108. the Pacific Ocean

109. a seismograph

110. Malcom

111. a glacier

112. fair weather

113. anemometer

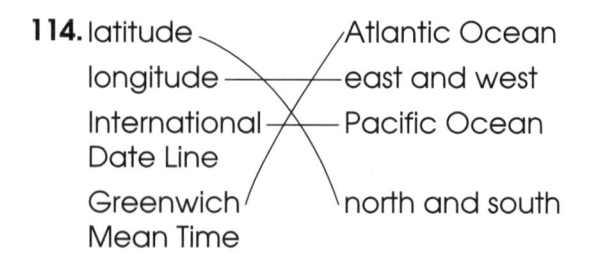

114. latitude — north and south
longitude — east and west
International Date Line — Pacific Ocean
Greenwich Mean Time — Atlantic Ocean

115. Philippines

116. Antarctica

117. (water,) plants, animals, (wind)

ANSWER KEY

118. 20,000 meters

119. the mantle

120. the Northern Hemisphere

121. Alaska

122. (limestone), quartz, (marble), granite

123. talc, shiny, magnetite

124. Crossword: SHELF, SLOPE, PLAIN, TRENCH

125. Antofagasta, Chile

126. B

127. (heat), glaciers, wind, water

128. mesosphere

129. D

130. sulphur dioxide

131. C

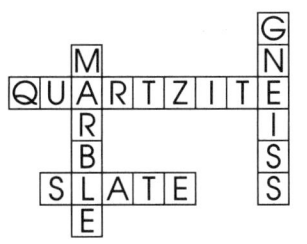

132. Crossword: QUARTZITE, MARBLE, SLATE, GNEISS

133. a hurricane

134. (Long Island), Hawaii, Iceland, Krakatoa

135. 8.0

136. F-2 tornado = 182–253 K.P.H. (113–157 M.P.H.); category 2 hurricane = 154–161 K.P.H. (96–100 M.P.H.)

137. sulphur dioxide, air, sewage

138. (fold), block, domed

139. 4643 meters (15,232 feet)

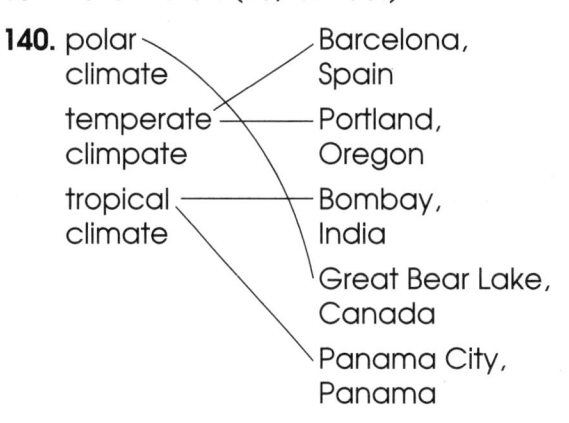

140. polar climate — Great Bear Lake, Canada; temperate climate — Barcelona, Spain; Portland, Oregon; tropical climate — Bombay, India; Panama City, Panama

141. 908 kilograms (2000 pounds)

142. Peru, Bolivia, (Panama), Chile

143. A

144. Different-sized jars have different diameters. The diameter of the jar will affect the amount of water that is measured. Also, some water could have evaporated during the course of the week.

SPACE SCIENCE

145. 40,000 grams, or 40 kg

146. Mercury

147. light

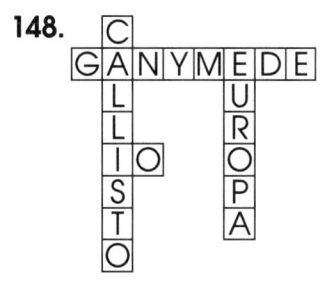

148. Crossword: GANYMEDE, CALLISTO, EUROPA

149. light-year, (magnitude), parsec, kilometer

61

ANSWER KEY

150. galaxy — Milky Way; constellation — Orion; moon — Oberon; star — Polaris

151. blue-white star

152. T, F, F

153. the Earth's core — 3550 km thick; the Earth's crust — 64 km thick; the Earth's mantle — 2900 km thick

154. 4

155. Saturn

156. 1/2 kg of cereal on Earth

157. 6 minutes, 40 seconds (approximately)

158. B

159. Neptune, Saturn, Jupiter, (Venus)

160. 1986

161. Neptune

162. Crossword: JUPITER, SATURN, NEPTUNE, URANUS

163. 4.3 years

164. A

165. the moon — light travels, sound does not travel; Earth — light travels, sound travels

166. comet, star, asteroid, (meteor)

167. constellation, galaxy, winter sky

168. C

169. galaxy, asteroid, (universe), solar system

170. 1095 days, or 3 years

171. Sound cannot travel on the moon.

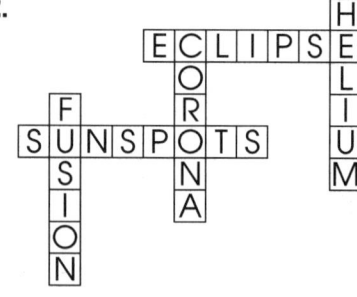

172. Crossword: ECLIPSE, HELIUM, SUNSPOTS, CORONA, FUSION

173. one-half

174. B

175. 680° C

176. the North Star (Polaris)

177. They are named after characters in Shakespeare's plays and Alexander Pope's poetry.

178. December 21 or 22

179. Rotate east to west

180. The Southern Hemisphere is tilted toward the sun.

181. B

182. F, T, F

183. Mount Maxwell

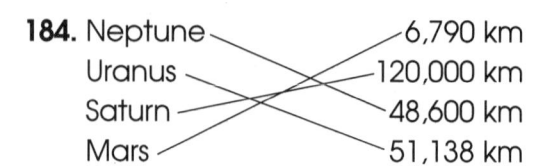

184. Neptune — 48,600 km; Uranus — 51,138 km; Saturn — 120,000 km; Mars — 6,790 km

185. approximately 52.1 kg

186. blue-white, blue, green, yellow, orange, red

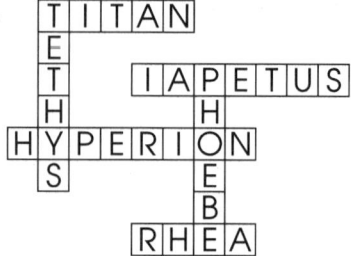

187. Crossword: TITAN, TETHYS, IAPETUS, HYPERION, PHOEBE, RHEA

188. That's the time it takes the Earth to make one complete rotation around its axis. It's also known as a day.

ANSWER KEY

189. on the planet Mars

190. A

191. Polaris (the North Star)

192. Saturn, Pluto, Earth, Uranus (all of them)

EXTENDED CHALLENGE

1. Categories with connections:

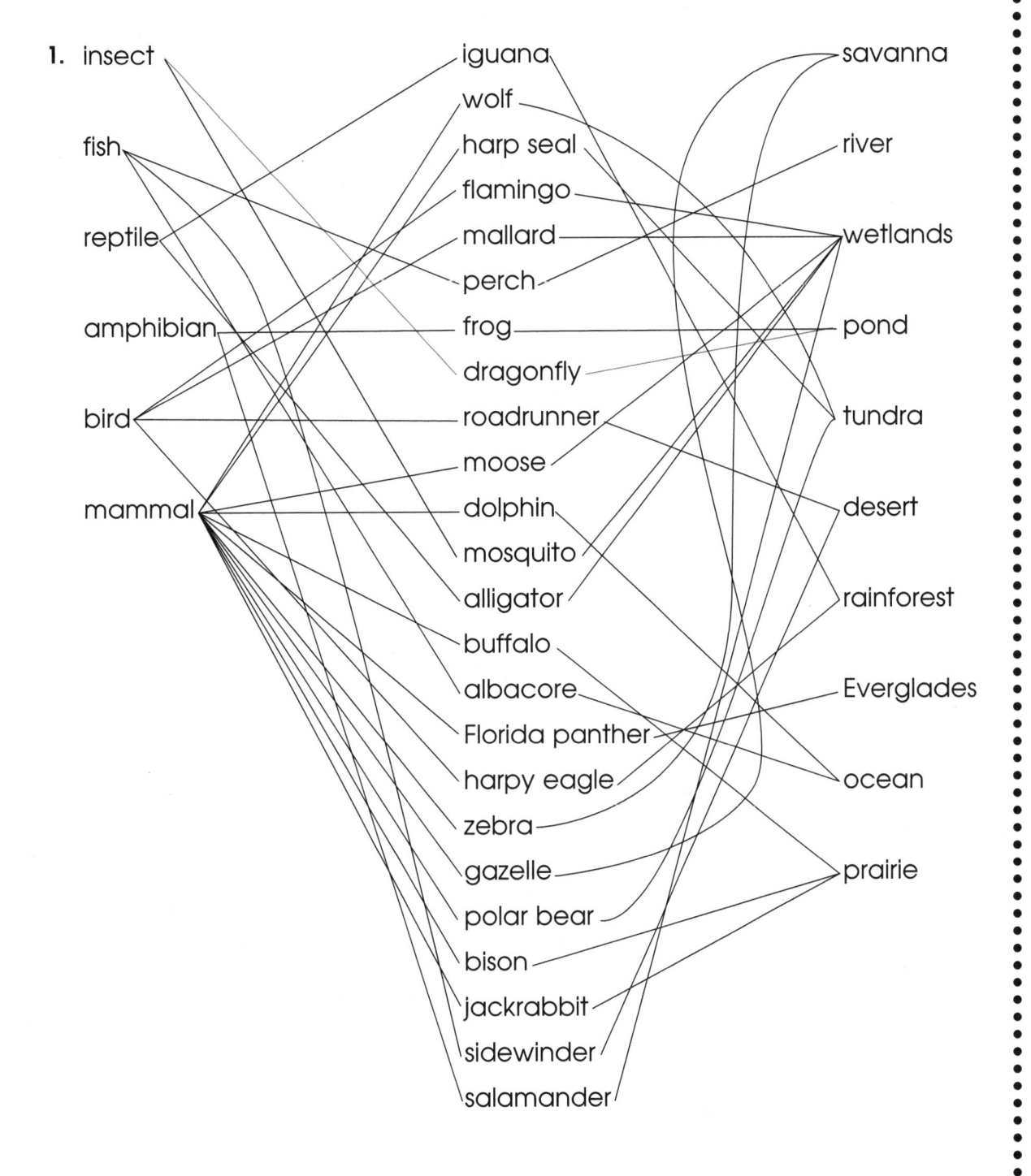

Animal types: insect, fish, reptile, amphibian, bird, mammal

Animals: iguana, wolf, harp seal, flamingo, mallard, perch, frog, dragonfly, roadrunner, moose, dolphin, mosquito, alligator, buffalo, albacore, Florida panther, harpy eagle, zebra, gazelle, polar bear, bison, jackrabbit, sidewinder, salamander

Habitats: savanna, river, wetlands, pond, tundra, desert, rainforest, Everglades, ocean, prairie

ANSWER KEY

	TRIASSIC	JURASSIC	CRETACEOUS
2. Allosaurus	—	X	—
Ankylosaurus	—	—	X
Apatosaurus	—	X	—
Brachiosaurus	—	X	—
Coelophysis	X	—	—
Compsognathus	—	X	—
Iguanodon	—	—	X
Oviraptor	—	—	X
Psittacosaurus	—	—	X
Rutiodon	X	—	—
Stegosaurus	—	X	—
Triceratops	—	—	X
Tyrannosaurus	—	—	X

3. A. digestive
   B. circulatory
   C. respiratory
   D. excretory
   E. nervous
   F. reproductive
   G. muscular
   H. skeletal
   I. sensory

   E brain   C trachea
   G tendon   F ovary
   C alveoli   D kidney
   A intestine   G triceps
   B aorta   A salivary gland
   F sperm   C diaphragm
   D ureter   I eardrums
   D sweat gland   H cartilage
   B vein   I pupils
   H tibia   B valve
   E nerves   I olfactory
   G ligaments   E axon
   A esophagus

From Science Challenge by Anthony D. Fredericks. Copyright © 1998 Good Year Books.

64

ANSWER KEY

4.

```
              M A N E D W O L F
            C H I M P A N Z E E
  C A L I F O R N I A C O N D O R
              H U M P B A C K W H A L E
                B E N G A L T I G E R
              O R A N G U T A N
      G I A N T O T T E R
                  B R O W N P E L I C A N
      G A L A P A G O S P E N G U I N
      G I A N T P A N D A
        F L O R I D A P A N T H E R
        A F R I C A N E L E P H A N T
      H A W K S B I L L T U R T L E
                M E X I C A N G R I Z Z L Y B E A R
        J A G U A R
              B L A C K R H I N O
      M O N K S E A L
```

5.

		FRICTION	MAGNETISM	ELECTRICITY	GRAVITY
1.	Ride a bicycle	yes	no	no	yes
2.	Ski down a hill	yes	no	no	yes
3.	Shine a flashlight	no	no	yes	no
4.	Operate a model train	yes	yes	yes	no
5.	Turn on a TV set	no	yes	yes	no
6.	Push a car out of the snow	yes	no	no	no
7.	Walk on the moon	yes	no	no	yes
8.	Make a telephone call	no	yes	yes	no
9.	Brush your teeth	yes	no	yes/no	no
10.	Use a computer	no	yes	yes	no

ANSWER KEY

6.

BICYCLE PART	SIMPLE MACHINE	CAN OPENER PART	SIMPLE MACHINE
wheels	wheel & axle	handles	lever
derailer	wheel & axle	turning key	wheel & axle
front sprocket	wheel & axle	cutting blade	inclined plane
gear wheels	pulley		
pedals	wheel & axle		
handlebars	wheel & axle		
brake handles	lever		

7. A. mechanical
    B. thermal
    C. chemical
    D. electrical
    E. nuclear

    __C__ gasoline
    __C__ battery
    __D__ light bulb
    __E__ atomic bomb
    __A__ bicycle
    __B, C, D__ oven
    __D__ computer
    __D__ generator
    __A__ pulley
    __B__ sun
    __A__ water wheel
    __B, D__ heating pad
    __C__ coal
    __B__ boiling water

8. 1. conduction
    2. radiation
    3. convection
    4. radiation
    5. convection
    6. conduction

ANSWER KEY

9.
```
W I N D C U R R E N T S S D Y
H R T U E L E C T R I C I T Y
U M W O X Q L P D L A S E B T
R W A V E S P I N I M Y E W H
R C B Y A O I W R G E T M N U
I Z C O P W S B M H A O E Y N
C T H U N D E R S T O R M S D
A R T Y U L V A E N P N V C E
N D F O E P E I M I B A B H R
E B L N R N P N R N D D W Q H
P C N W O H J K L G E O W S E
U E L D E P R E S S I O N A
F W C E X P L O S I O N W S D
P Y A S W H A I L S T O N E S
C M S T R O N G W I N D S S S
```

The name of the storm missing from the puzzle is <u>typhoon</u>.

10.

	FOLDED	FAULT-BLOCK	DOMED	VOLCANIC
1. Sierra Nevadas	___	X	___	___
2. Black Hills	___	___	X	___
3. Rocky Mountains	X	X	___	___
4. Appalachians	X	___	___	___
5. Cascades	___	___	___	X
6. Wasatch Range	___	X	___	___
7. Tetons	___	X	___	___
8. Ozarks	___	___	X	___

67

ANSWER KEY

11.
1. Arctic Ocean
2. Pacific Ocean
3. Caribbean Sea
4. Red Sea
5. Indian Ocean
6. Gulf of Mexico
7. Mediterranean Sea
8. Hudson Bay

12.

KIND OF ROCK	MAJOR CLASSIFICATION	CHARACTERISTICS	HOW IT WAS FORMED
granite	igneous	large crystals	slowly cooling lava
shale	sedimentary	fine-grained	sediments of mud/clay/silt
quartzite	metamorphic	very hard; originally sandstone	great heat; great pressure
gneiss	igneous	salt & pepper appearance	slowly cooling magma, then great pressure
sandstone	sedimentary	grainy	sediments cemented over time
pumice	igneous	gray, floats on water	rapidly cooling lava
diamond	metamorphic	hardest mineral	great pressure
marble	metamorphic	white	great heat & pressure
limestone	sedimentary	chalky	sediments of sand built up over time
obsidian	igneous	black, glassy	rapidly cooling lava

ANSWER KEY

13.
1. F	6. F	11. T
2. T	7. F	12. F
3. F	8. T	13. F
4. T	9. F	14. T
5. F	10. T	15. F

14.

AVERAGE DISTANCE FROM SUN (IN MILLIONS OF KM)	PLANET	LENGTH OF DAY (IN EARTH TIME)
4497	Mercury	116.7 days
1425	Venus	24.6 hours
228	Earth	10.4 hours
108	Mars	18.5 hours
5900	Jupiter	6.4 days
2867	Saturn	176 days
778	Uranus	24 hours
150	Neptune	9.9 hours
58	Pluto	16 hours

ANSWER KEY

15.

CONSTELLATION	POPULAR NAME	BRIGHTEST STAR	BEST SEEN DURING
Taurus	Bull	Aldebaran	winter
Auriga	Chariot Driver	Capella	fall
Cepheus	King	none	all seasons
Orion	Mighty Hunter	Betelgeuse	winter
Ursa Minor	Little Dipper	Polaris	all seasons
Canis Major	Orion's Dog	Sirius	winter
Cassiopeia	Queen	none	all seasons
Scorpius	Scorpion	Antares	summer
Leo	Lion	Regulus	spring

16.

```
  M E R C U R Y
    M E T E O R
    P L U T O
  S A T E L L I T E
V E N U S
    S T A R S
J U P I T E R
    E A R T H
  P L A N E T S
    B L A C K D W A R F
    M O O N
    R E D G I A N T
A S T E R O I D S
  C O M E T S
  M A R S
```

70